I0463381

Amazon Echo Dot

Advanced User Guide - Step by Step Instructions to Enrich Your Smart Home

By: Matthew Johnson

breached.

Disclaimer Notice:

Please note the information contained within this document is for educational and entertainment purposes only. Every attempt has been made to provide accurate, up to date and reliable complete information. No warranties of any kind are expressed or implied. Readers acknowledge that the author is not engaging in the rendering of legal, financial, medical or professional advice.

By reading this document, the reader agrees that under no circumstances are we responsible for any losses, direct or indirect, which are incurred as a result of the use of information contained within this document, including, but not limited to, — errors, omissions, or inaccuracies.

TABLE OF CONTENTS

Amazon Echo Dot

Introduction

Welcome to the world of the Amazon Echo gadgets and the mystery that is Alexa. 2014 saw the introduction of the Amazon Echo to a star struck world that immediately proceeded to purchase the device in droves. The Amazon Echo Dot, launched in 2016, was no less successful. So, successful were these devices, in fact, that after the Thanksgiving sales, it was discovered that all the devices being sold online were sold out! This gave rise to the rumor that the Echo Dot had been discontinued - a rumor that was quickly debunked by a justifiably smug Amazon. Google, Apple and Microsoft have been racing to button up, and there is plenty of speculation about who will win this battle of the voice assistants – Google Assistant, Alexa, Cortana, or Siri.

But while this speculation rages on, chances are that you have either received or purchased an Amazon Echo Dot. Congratulations! This eBook contains a guide to what the Echo and Echo Dot are and how you can install these devices. It also contains a detailed look at Alexa, her skills, how you can use her most efficiently, and some light-hearted Easter eggs.

I have attempted to give you a thorough look at the Amazon Echo and Echo Dot and have even made some recommendations about some of the things you should purchase with the device.

I hope this book is helpful to you and assists you in getting the most out of your Amazon Echo Dot. Happy reading!

Chapter 1: A Look at Voice Assistants and Their Growing Influence

An intelligent personal assistant (IPA) is a product specialist that can perform services or errands for a person. These services or assignments depend on client input, area mindfulness, and the capacity to get to data from an assortment of online sources (weather conditions, news, stock costs, traffic, client plans, retail costs, etc.). Examples of such operators are Google Home, Apple's Siri, Google Now (and later Google Assistant), Amazon's Evi (marketed as Evi), Amazon Alexa, the open-source, Lucida, Microsoft's Cortana, Samsung's S Voice, Braina (an application created by Brainasoft for Microsoft Windows), BlackBerry's Assistant, LG G3's Voice Mate, SILVIA, IBM's Watson (PC), HTC's Hidi, One Voice Technologies (IVAN), and Facebook's M (application).

Examples of jobs that an Intelligent Automated Assistant might perform include those of a personal assistant: timetable administration (e.g., sending a message to a dinner date that a client is running late because of traffic, overhauling plans for both sides, and changing the restaurant reservation time), and personal well-being administration (e.g., observing caloric needs, heart rate, and exercise regimen, then settling on proposals for future improvements).

Intelligent personal assistant innovations are empowered by the mix of cell phones, application-programming interfaces (APIs), and the multiplication of versatile applications. Be that as it may, intelligent robotized assistants are intended to perform particular, one-time assignments indicated by client voice directions, while savvy personal operators perform progressing errands (e.g., plan administration) on their own.

One of the key parts of an intelligent personal assistant is its capacity to sort out and look after data. This incorporates the administration of messages, scheduling events, the management of documents, and so forth.

Contrasting an automated assistant (AI) with a personal assistant, as many organizations have done recently, makes for a useful comparison. It is one that is demonstrative of the human abilities that most significant innovation organizations need their immaterial assistants to embrace. In the course of the last couple of years, with enhancements in discourse acknowledgment innovation, Cortana, Siri, and Google Now have gradually figured out how to move past the nuts and bolts of climate overhauls to go up against more unpredictable duties, like dealing with your calendar or noting your questions. In any case, items that deal with our personal spaces - like Amazon's Echo and Google Home - indicate that a bigger development in human-gadget communication is presently in progress.

Audience demos of Google Home, which has the organization's Assistant incorporated with it, show a conversational capacity that requires a superior comprehension of human context and intent. The gadget depends altogether on the organization's discourse acknowledgment innovation that has been taking shape for approximately ten years- since the beginning of GOOG 411. Be that as it may, throughout the years, the essential phone-based registry look has developed into the considerably more mind-boggling Google Now.

Amazon's Echo's biological system depends on virtual assistant Alexa to react to voice instructions.

The exceptional leap in the Android assistant's abilities originated from neural net preparing and profound learning strategies that have permitted researchers to improve discourse acknowledgment innovation to a point where it is presently beginning to take in the subtleties of human conduct through the medium of voice.

Utilizing the voice to speak with an outside element makes for a personal and inherently human experience. "Discourse is the most prevailing way that mankind has been speaking with each other," David Nahamoo, discourse CTO at IBM Research, said via telephone. "When we speak with the outside, we talk. Be that as it may, from outside to inside, we retain data a great deal better outwardly. This is a direct result of our legacy and the advancement that we have experienced. From the angle of productivity, discourse is the fastest approach to get a point across."

Voice changes the way individuals connect with their frameworks. Google Gadgets like Echo and Google Home, for example, are based on discourse acknowledgment that can help you stay hands-free and up-to-date while you do multiple other things around the house. So, as opposed to investing energy in

12

swiping or writing, you can tell the personal assistant what you require or what you are searching for. It's that sort of straightforwardness and efficiency that organizations dangle before their clients to have them embracing chatbots and personal assistants in their day-to-day correspondences. Yet conversing with gadgets likewise opens the way to another sort of relationship.

While a verbal trade with a virtual assistant can make the completion of tasks less demanding, it additionally makes it easier for organizations to increase the amount of priceless knowledge the virtual assistant has about the human world through vocal pieces of information about an individual's sentiments and inclinations. It is clear that we are moving toward virtual assistant comprehension, whereas earlier we focused on calculation. This is not only about us talking. These technologies are taking note of what we like and do not like, how we talk, and what we talk about. They now have great access to how we think.

Within the realm of AI, information is the means of separating one organization from the other. Through vocal searches, a large number of vocal specimens become distinctly accessible to the organizations that are calibrating personal assistants. The flood of data is fed again into the framework to enhance the exactness of the calculations, yet it additionally gives the organizations access to the complexities of human aim. As a result, utilizing the voice to speak with an AI aide just makes it smarter and more adaptable.

A great deal can be gathered from vocal correspondence. Words and pitches begin to give away client inclinations and even feelings after some time. That sort of insight into the attitude of the client is essential to the following rush of personalized AIs from organizations like Facebook, Amazon and Google.

Savvy, talking AIs at home will create a new environment, making themselves a necessity instead of the curiosity they are today. With organizations seeking to make their assistants ubiquitous and their machines more interconnectable, they require competent recognition of speech to take care of business.

There is a parallel push. It will become normal to connect with your tech-savvy cooler or printer via voice activation. Additionally, you could have a convenient personal assistant that lives in a cloud and chases after you to help you explore a perplexing world. Google Home, much like Amazon's Echo, now connects to items that are valuable around the house. You can utilize the speaker to control your Chromecast, Nest, and Philips Hue lights.

Notwithstanding quickly exploring the physical world, an ever-present assistant could conceivably become an entryway to new settings or remote dialects as well. In the spot disclosed amid the Google event this week, the organization exhibited that Home can tap into Google Translate to react with precise interpretations from English to Spanish. Whether the machine can appreciate and decipher outside accents remains to be seen.

These machines are listening to what we like and do not like, how we talk, and what we talk about. It is more noteworthy that they can access how we think. Neglecting to appreciate diverse accents has been one of the greatest disgraces of most advanced assistants on cell phones today. Researchers constructing these frameworks regularly discuss the absence of information as one of the greatest hindrances to deciphering new accents and dialects. The measures of data required to make those conceivable calls for great speculation from the organizations. Taking the innovation straight to individuals' homes opens up a constant flow of information that can be utilized for tests back in the exploration labs.

A lot of the basics have fallen into place for gadgets such as Google Home to end up as distinctly effective personal assistants. What's more, despite the fact that we should be careful of the ways human-gadget associations are beginning to move, most voice interface designers trust that it is a fundamental change that will amplify human capacities.

Having an AI that is your operator helps you exist on the planet better, and improves data and administrations which is enormously energizing. Of course, as with any new technology, there are consequences that can be negative. We are all acquainted with security and mining information concerns. That is something we must be attentive about, yet the advantages far exceed the potential danger of situations.

Chapter 2: Amazon Echo and Amazon Echo Dot

Amazon Echo (referred to as Project D, or Doppler, while being developed, and abbreviated and alluded to as Echo) is a smart speaker (along the lines of a smartphone or smart TV) created by Amazon.com. The gadget comprises of a 9.25-inch (23.5 cm) tall chamber speaker with a seven-piece arrangement of microphones. The gadget interfaces with the voice-controlled, intelligent personal assistant, Alexa, which reacts to the name, "Alexa." The "wake word" to bring Alexa to life can be changed by the owner to either "Amazon" or "Echo."

The gadget is capable of voice connection, music playback, making schedules, setting alerts, streaming videos and podcasts, playing book recordings, and giving traffic, weather, or other synchronized data. It can likewise control a few smart gadgets, utilizing itself as a center point for the automatization of your home.

Various reports affirm that Amazon has been creating Echo inside its Lab126 workplaces in Cambridge, Massachusetts and Silicon Valley since 2010. The gadget was a piece of Amazon's first endeavors to grow its gadget portfolio past the Kindle tablet. The Echo was unmistakably highlighted in Amazon's first Super Bowl advertisement in 2016.

Echo use was at first limited to Amazon Prime customers and others who were specifically invited. On June 23, 2015, it was made generally accessible in the United States. Reporters guessed that it would make its Canadian presentation sometime in the latter half of 2016, after Amazon co-facilitated a hackathon in Toronto and began a job search for engineers for Alexa. The Echo became accessible in the United Kingdom on September 28, 2016. Moreover, the Alexa voice administration can be added to different gadgets, and other organizations' gadgets and services are urged to interface with it.

In March 2016, Amazon revealed the Amazon Echo Dot, which is a version of the Echo that is the size of a hockey puck, and that is intended to connect to existing speakers. The Amazon Echo Dot permits you to utilize all the same abilities of the Amazon Echo, except in a much smaller, and more portable size.

Amazon made Alexa accessible as a web application for testing purposes in May of 2016, portraying it as a reenactment of the look and feel of the Amazon Echo. The test system was meant for engineers outside of the United States, where Echo was not yet accessible.

The second form of the Amazon Dot became available on October 20, 2016. It highlighted a lower cost and enhanced voice acknowledgment, as well as offered two colors – white and black. The new Echo Spatial Perception (ESP) innovation permitted many Dot and Echo units to cooperate, so that just a single gadget answered the demand. With its 360-degree speaker, the Amazon Dot utilized far-field voice acknowledgment, fueled by the Alexa voice assistant. In terms of design, it had the appearance of a hockey puck with two buttons on top. Fundamentally, it is a Bluetooth speaker that resembles the top part of a Pringles can.

Everything the old Echo had could now be found in the Echo Dot, including long-range, far-field voice capacities, a button to switch off the microphone, a button for action (for Wi-Fi), and, obviously, "Alexa," who can play music, answer questions, and give you news, traffic, weather, sports results, and a great deal more. Even the Echo's outside abilities, like Uber, Fitbit, Capitol One (found in the Alexa application), and Domino's Pizza, remain accessible.

New Features & Hardware of the Echo Dot

Although the Echo Dot software is like its forerunner's, the equipment is somewhat unique. That is the thing that makes it intriguing. This item has a 3.5 mm sound yield jack that allows it to network to all speakers and transforms it into a gadget that is powered by Alexa. You will also find that a power connector, a 3.5 mm sound link that is 4 ft. long, and a USB charging link (connector and USB link included with Amazon Echo) are included in the package that you receive.

You may even interface the gadget with your own Bluetooth speaker, if you would like to. Once you download the Alexa application from Android, Fire OS or iOS gadgets, you can utilize the included Quick Start Guide to setup your Wi-Fi and Bluetooth to other devices (Bluetooth speakers and cell phones). The absence of an output for audio on the main Echo generated many complaints when it was released. Minor changes like this one have a major impact.

You can get the sound quality you need and access Alexa's capabilities, including the ESP highlight intended for different gadgets. When you say "Alexa," just the nearest gadget to you will react. The Echo Dot is much smaller and lighter than the Echo, a fact that expands its versatility in the home. You can even take the Dot outside your home if you have connected it with another device powered by Alexa, the Amazon Tap. However, the Amazon Tap is seen as an average Bluetooth speaker that lacks cell information as well as versatile hotspot associations, accordingly restricting its mobility. The Tap might be battery fueled, however when you always have a Wi-Fi connection, it is not up to the mark.

The Echo Dot, similarly to the Echo, has a wake word that allows for hands free interaction. The wake word that initiates the gadget can be "Alexa," "Echo," or "Amazon." Because it is battery controlled, you need to "tap" button the gadget for Alexa to react to your query or demand.

The major factor that sets the Amazon Echo Dot apart from the Amazon Echo is that the Echo Dot is half the price, selling for $49.99. With access to all the original features of the Amazon Echo, enhanced speakers, and a lower cost, it is no surprise that a large number of tech websites commended this item, frequently giving it an Editor's Choice Award.

You Do not Need an Extra Speaker

What I really like about the Echo Dot is that it does not lose the distant voice acknowledgment of its precursor. The Echo Dot has an arrangement of seven microphones that allows for simple and quick Wi-Fi connectivity, and, above all, has a speaker that is built-in and that conveys sound comparable to other, smaller Bluetooth speakers. The sound quality is not the greatest and will not make your room resonate with music, yet it is not bad if you want to listen to it in the intimacy of your bedroom. Numerous commentators agreed that it sounded better than anticipated.

How to Order an Echo Dot

At the time the original model was launched, there were only three ways to order it:

- Voice-controlled Amazon Fire TV

- Amazon Echo

- Amazon Tap

The individuals who pre-requested the primary model a month prior to its release (March 31, 2016) were lucky to get it early, however there were restrictions on the number of devices you could buy. Amazon prefers not to give out their purchase statistics, however, you can assume the device sold well based on the success of the first generation.

You can now place an order for the second-generation version on the web or in-store. You no longer need to buy using Alexa (in spite of the fact that if you do so you have the enticement of a $10 shopping credit). Also, there are no more limits on the number of devices per client or order. You can now place an order for however many devices you need. In fact, if you place an order for five devices, you get one device free.

It is not surprising that this amazing item at half the cost of previous models is Amazon's number one seller. Watch out for unique Black Friday deals and packages from Amazon, since you may get the opportunity to buy the Dot for just $39.99, or in a package that incorporates a Philips Hue starter unit. Make certain to look at Amazon.com for bargains all through November and December.

Chapter 3: An Introduction to Alexa

Alexa, whose name is derived from the ancient library that was called Alexandria, is Amazon's voice control framework. It gives you a chance to voice requests, such as switching off your lights or playing music track, and have them met. It turns the Echo, Dot, and Tap streaming sound players into Internet assistants and true smart home centers. Also, it is what transformed Echo into not only one of the year's greatest tech items, but potentially the greatest development so far for those intrigued by brilliant home control.

Because of the Echo's far-field receivers, Alexa can react to voice orders from any place that is within earshot. Furthermore, you do not have to press any button to activate. You simply say the trigger word, or the wake word (either "Echo," "Alexa," or "Amazon"), after which you state your request Alexa will then fulfill your request, provided that you have set up everything correctly and are utilizing the right instructions (remember that it is still being developed and improved, and you need to make your demands in accordance with that). When you get used to the idiosyncrasies, utilizing Alexa feels considerably more normal than addressing a telephone-based voice assistant, like Apple's Siri. Accordingly, you will likely wind up utilizing your telephone less when you are home.

There are some privacy concerns regarding the way Alexa is continually tuning in, and there is still a great deal it cannot do. However, there are enough benefits to make the Echo an appealing item, particularly if you are interested in smart home technology.

Who ought to get an Echo Dot?

Let's be honest: though the Echo is a respectable speaker in its own right, you can improve sound for less cash if music is all you are after. The real motivation to get an Echo at this moment is the Alexa voice control system. Voice control liberates you from being continually dependent upon your cell phone. In case you are one of those people who likes gadgets like cell phones that switch lights on and off, Alexa affords you the opportunity to nerd out completely as she can control all of your gadgets.

Alexa is especially significant for brilliant home clients, since it gives you the ability to control your associated gadgets without taking out your telephone and using an application. (Alexa has an application, however that is primarily for setup and design, or to include new capacities or successions of summons, called "aptitudes," or to view to-do and shopping records). More often than not, the Echo, Tap, and Dot let you access the vast majority of Alexa's helpful elements without requiring you to interact with a screen by any means.

Envision strolling into your home at night with your arms filled with supplies. To turn on the lights you would have to put down the bags, pull out your cell phone, open it, find the application, identify the control for the lights you need, and tap the symbol. With Alexa, you simply speak the words "Alexa, turn on the kitchen lights." Presto! The lights turn on.

Past requesting music, you can utilize Alexa to look at Wikipedia ("Alexa, who was Abraham Lincoln?"), make speedy cooking calculations ("Alexa, how many pints are in a gallon?"), help with math homework ("Alexa, what's 48 x 9?"), or make a schedule ("Alexa, include 'go to the vet' in my to-do list."). The great thing about your Alexa gadget is that it continues enhancing the more you possess it.

On the off chance that you, as of now, have some Alexa-ready gadgets, or one of the three noteworthy, brilliant home center points (Wink, SmartThings, and Insteon), adding an Echo can make using those gadgets all the more fascinating and effective.

In any case, Echo is not a far-reaching, brilliant home framework. It will not supplant a customized home center, nor will it take into account complex combinations of different gadgets ("turn the lights green when dance music plays in the kitchen"), or generally be the most viable method for connecting with your different gadgets. But it is a useful interface for your home that performs in a way that an application on your telephone cannot. Therefore, you will likely increasingly come to depend on it.

How Alexa functions

Alexa is one of the easiest programs to interact with. When you pose a question, or convey an order to Alexa, you generally do not need to ask twice. A portion of Alexa's success is due to the seven exceptionally delicate mouthpieces incorporated with both the Echo and Dot (the Tap, being a compact adaptation, works differently). Alexa is continually tuning in, and rushes to react.

Echo and Alexa are for people who want a simple device interface, so they can focus on enjoyment, rather than on programming. As noted before, Echo and Dot stream your voice to the cloud when you make a request (your solicitations additionally appear in the application, and you can erase them on the off chance that you would prefer not to leave a record). The default wake word is "Alexa," but you can change that to "Echo," or "Amazon." Once you say the magic word, the amplifier takes what you say next and sends it up to Amazon's cloud PCs for quick processing. If you ask a question ("Alexa, what's the climate in Chicago today?") a computer-generated, female voice answers. If you want music, Alexa will look through the Amazon Music inventory or Amazon Prime Music to fulfill the request. (An Amazon Prime account is required for access to Prime Music.) It can likewise play TuneIn Internet radio stations, music from Pandora (free or paid accounts), Spotify (just supporter records), and iHeartRadio (membership required). If you request a joke, be prepared to moan. A request to turn off a light or change the temperature will result in immediate action.

It's those brilliant home-regulating abilities that have many individuals excited about Echo. One of its greatest appeals is the ability to make normal, day-to-day errands simpler. While killing the lights or changing the temperature with an application on your iPhone is slick, finishing a similar assignment just by speaking is considerably less demanding.

Echo currently has a short list of items it can easily interact with. That list incorporates the most well-known smart home items available, and is constantly developing. The list of items Alexa can work with is supplemented by something Amazon calls "aptitudes," and outsider incorporation applications like IFTTT, Yonomi, and Muzzley (more on those underneath). Amazon invites more incorporation applications, and its API is accessible to any organization that wants it.

The least demanding combinations are with the items Alexa can work with directly, and those can be found in the smart home area of the application's menu. Application frameworks incorporate Philips Hue lights, the Lutron Caséta lighting framework, Belkin WeMo switches (both in-divider and module, however no other WeMo items), Nest indoor regulators (second-and third-era models), Ecobee indoor regulators, the Sensi indoor regulator, the TP-Link Kasa outlet switch, and LIFX savvy globules, in addition to Insteon, Wink, and SmartThings center points. Coordinating with one of the center points makes Echo ready to control most things associated with those centers.

Another key component of Echo Dot and Alexa, and one that is near Amazon's heart, is voice-controlled shopping. To purchase something, you request that Alexa search for your desired item. It will search, let you know the cost, and request that you affirm with a four-digit security code (that you would have already recorded within the application). This last feature keeps kids and outsiders from making unapproved purchases. Once authorized, the purchase is charged to your default payment method.

Alexa's Skills

What's an application? In the Echo/Alexa world, an application resembles an application. It is a program you can add to your Echo/Dot/Tap to empower some new abilities. As this book is being written, there are around 1,500 aptitudes in the application menu of the Alexa application, with there are addition made each week. Toward the beginning of April, Amazon made the Alexa Skill Kit available to the general public, making it simple for anybody with basic programming abilities to add applications to Alexa. When you add an application to your Alexa account, it will work with all the Alexa gadgets in your home. You do not have to add the ability to every gadget independently.

Like applications for cell phones, there are great deals of disposable applications; however, there are a lot of valuable ones also. The main application that any smart home devotee will need to include is IFTTT application (If This Then That). IFTTT is an application that allows you to connect diverse gadgets in the cloud with formulas.

A few applications let you buy items or administrations outside of Amazon. For example, there is an Uber application for requesting a ride (there is also a Lyft ability for the same), a Domino's Pizza application for ordering pizza, and an application for requesting flowers from 1-800-FLOWERS. That's just the beginning.

Clients can find applications for an assortment of leisure activities and interests. There are tide-finders for surfers, homework aides for students, formula programs, and activity report functions, in addition to many applications for individual smart home gadgets. For example, there is the SkyBell doorbell camera and the Scout smart home security framework.

Among the less practical (but lots of fun) applications are DrinkBoy (for discovering drink formulas), Cricket Facts (for facts about crickets), Daily Affirmation (to improve you feel about life), unicorn trivia, and no less than two abilities to deliver "yo mom" jokes. There is no limit to the number of abilities you can add to your account, but if you have quite a few, getting them to interact well together can be difficult.

Alexa's impediments

One of the biggest issues Alexa users have—even users who are enthusiasts of the framework—is that it can only handle one request at a time. You cannot state, "Alexa, turn on the parlor and living room lights," unless you have already made a group called, "Parlor and Living Room." You cannot state, "Alexa, play the Beatles, and turn up air." Each order must be given independently. Regardless of how many gadgets you have linked to Alexa, you can only address one at a time.

Correspondingly, you can make a shopping list by advising Alexa to add sparklers or brew to your list, however you should include everything independently. For instance, rather than saying "Alexa, add fireworks and brew to my shopping list," you need to state, "Alexa, add sparklers to my shopping list. Alexa, add brew to my shopping list." If you are making a substantial shopping list, the procedure can get irritating, particularly for other individuals in the room. However, it is easy to make a shopping list in the Alexa application, and you can access the list when you are in the grocery store. To share or print the list, you would have to utilize an IFTTT formula.

What's more, though users love to ask Echo arbitrary questions, Alexa is not Google. The framework knows a lot, however what it cannot answer exceeds what it can.

The far-field receivers on the Echo and Dot are sensitive to the point that, on the off-chance that I address the Dot in my basement, and still have the door open, the Echo in the kitchen will probably still pick up on it.

Owners can change the wake word for each Alexa gadget, however it gets confusing to remember which one reacts to "Alexa" and which one reacts to "Echo." Amazon plans to address this issue with the presentation of its new Echo Spatial Perception innovation, which actuates voice charges from the Echo gadget nearest to you.

Now and then, Alexa can not hear your request because of

34

outside interference. The gadget additionally can not tell who is talking or what number of individuals are talking.

Amazon sends programmed firmware overhauls every once in a while, and proprietors have announced execution changes after those redesigns. Some of the time, though not always, those upgrades are enhancements, but you cannot control when they arrive.

In the event that you utilize your Echo or Dot basically for music, and you are seeking an entire house music framework like Sonos, you ought to stay with Sonos. In spite of the fact that Echos, Dots, and Taps will give you a chance to play music in each room, they can not synchronize their music the way a genuine house music framework can, in light of the fact that the stream to every speaker is free. You can not gather and control a cluster of Echos together.

You can not state, "Play The Misfits on every one of my speakers." You need to go to each Alexa gadget and instruct it to play music independently. Likewise, Amazon Prime's music rules let you play just a single stream at time, so in case you are playing Prime Music in the front room, you will have to utilize

Pandora or Spotify for music in the kitchen. Furthermore, on the off chance that you utilize TuneIn for podcasts, Alexa may disappoint. Attempting to locate a specific podcast or scene by means of voice hunt can be baffling. TuneIn stations may likewise turn out to be briefly inaccessible.

Is Echo keeping an eye on you?

Like any site or program, Alexa gathers data about how clients connect with it. A few people may worry about entrusting their money and personal data to a cloud-empowered voice assistant. Amazon compares its data-gathering to how sites monitor individual browsing habits for the purposes of marketing. For example, based on what of what you instruct it to do, Echo recognizes what music you listen to, what you put on your shopping list, and what brilliant home items you have associated with your framework. Probably that data is utilized to market more items to you, yet I would say utilizing Alexa consistently hasn't brought about more straightforward promoting from Amazon. Or possibly the association between my Alexa charges hasn't been as evident as, say, the stalking abilities incorporated with Chrome or other web programs.

Still, Alexa can likewise encourage correspondence with outside administrations, including the previously mentioned Uber and Domino's, and Capital One's web- based account administrations. Be that as it may, Amazon does not really access the exchanges occurring. In the example of Capital One, the bank says the framework is completely scrambled, and the Alexa aptitude incorporates a client-made password to avoid unapproved use.

The truth of the matter is that your Echo or Dot is continually listening to you. This is and is not as frightening as it sounds. In spite of the fact that the gadget hears all that you say inside the scope of its great far-field receivers, it's listening for its wake word. When it hears that, everything in the following few moments after is seen as an order and sent up to Amazon's Cloud where the right reaction or activity is activated. You know Echo is focusing when the round blue light turns on as it hears its name. Echo resembles your puppy: It's continually tuning in. However, it only focuses on key words such as "treat," "walk," or "mate." Everything else goes directly over its head.

This is the same as Apple's Siri and some of Samsung's smart TVs that also listen for key expressions such as "Howdy, TV" or "Hello, Siri". These devices respond to inquiries directed to them or to voice commands for things like volume and channel. Similarly, the Alexa gadgets only become functional when they hear their name (they additionally record a "small amount of a moment of sound before the wake word," as indicated by Amazon's Alexa FAQ page). So, when Alexa hears an order and sends those words up to the Cloud, Amazon does learn something about you. Possibly the organization might discover that you like The Police, or fart jokes, or need to put broccoli on your shopping list. In the event that you state, "Alexa, where should I cover the body?" you are not going to have the police appear at your entryway. (I know since I've attempted it).

Is it essential that Amazon gather this data? That is dependent upon you. Your PC is following all that you do online. Google knows all that you have ever searched for. Fundamentally, because it is a business, Amazon needs to offer you bunch of stuff. So when you utilize application your Echo to locate someone with workout application, do not be astonished if Amazon sends you an email promoting yoga pants. When you add mechanical pencils to your Alexa shopping list, Amazon may well utilize that to prescribe lead refills, or erasers.

Concerns about privacy have been raised both because of the

risks of information collection and the fact that Alexa is always listening. Would it be advisable for you to tell visitors that a recording gadget is listening to their discussions (comparatively, with a surveillance camera, do you tell your visitors that they're being recorded)? As far as voice commands are concerned, there is no difference between users. Anybody within range can utilize it, including kids. All that you say to Alexa is noted in the application and can be erased. Amazon says that once you erase it, it's gone permanently, even from its servers.

So when all is said and done, you are not prone to danger by using Alexa any more than you are with other similar services. Be that as it may, in the event that you want to talk about any national security concerns and would prefer not to open yourself to Alexa's snooping abilities, press the button on top to mute the microphone.

Chapter 4: Step by Step Instructions to Install Amazon Echo Dot

You can put Echo Dot in an assortment of areas, including your kitchen counter, your lounge room, your end table, or any place you need a voice-controlled PC. Echo Dot can be utilized with or without other Alexa gadgets.

Before you start utilizing your Echo Dot and the Alexa Voice Service, it needs to be connected to a Wi-Fi system. After that, you will need to use the Alexa app to register the device to your Amazon account. Follow the instructions below to do this:

1. Download the Alexa application and sign in

With the free Alexa application, you can set up your gadget, deal with your alerts, music, shopping records, and the sky is the limit from there. The Alexa application is accessible on telephones and tablets with:

- Android 4.4 or higher

- Fire OS 3.0 or higher

- iOS 8.0 or higher

To download the Alexa application, go to the app store on your cell phone and look for "Alexa application." Then select and download the application. You can also choose one of the following connections:

- Apple App Store

- Google Play

- Amazon Appstore

Likewise, you can go to https://alexa.amazon.com from Chrome, Safari, Microsoft Edge, Firefox or Internet Explorer10 or higher on your Wi-Fi-powered PC.

2. Turn on Echo Dot

Put your Echo Dot in a focal area (no less than eight inches from any windows and walls). At that point, connect the included power connector to Echo Dot, then connect the device to an electrical outlet. The light ring on Echo Dot turns blue, followed by orange. At the point when the light turns orange, Alexa welcomes you.

Note: Other USB control connectors, such as mobile phone chargers, may not be sufficient to power Echo Dot.

3. Associate Echo Dot with a Wi-Fi Network

Follow the guidelines in the Alexa application to interface Echo Dot with a Wi-Fi network. To take in more, go to the help section titled Connect Echo Dot to Wi-Fi.

Tips:

In the event that the set-up procedure does not immediately begin, press and hold down on the Action button that is on the Echo Dot for five seconds. At that point, open the Alexa application and go to Settings > Set up another gadget.

In the event that Echo Dot does not register your Wi-Fi, restart Echo Dot by unplugging and connecting the gadget once again.

On the off chance that despite everything you experience further difficulties, reset your Echo Dot to its factory settings and set it up again. For second generation Echo Dots, you can utilize the Reset Your Echo function.

4. Converse with Alexa

You can now utilize your Echo gadget. To begin, say the "wake word" Alexa and after that simply talk to the device. Your Echo gadget is set to react to the wake word "Alexa", but you can utilize the Alexa application to change the wake word whenever you would like. In the application, go to Settings, select your Echo gadget, and select wake word.

5. Connect Echo Dot to an External Speaker

While Echo Dot has a built-in speaker, you can use the included sound link or Bluetooth to connect it to outer speakers.

6. Associate Echo Dot to Bluetooth Speakers

You can also listen to unlimited streaming music from the Echo Dot by using another Bluetooth speaker.

Before you start:

- Place your Bluetooth speaker and Echo Dot no less than three feet apart. If Echo Dot is too close to the outer speakers, Alexa may have difficulty hearing either the wake word or other commands.

- For the best outcome, utilize a Bluetooth speaker confirmed for use with Echo Dot. For more data, navigate to Certified Speakers/Supported Bluetooth Devices for the Echo Dot.

- Ensure you can associate your speaker to other Bluetooth gadgets, similar to a cell phone.

- Turn on your Bluetooth speaker and increase the volume.

- Disconnect any other Bluetooth gadgets from Echo Dot. Remember that Echo Dot only associate with one Bluetooth gadget at any given time.

How To Associate Echo Dot With a Bluetooth Speaker

- Turn on *matching mode* on your Bluetooth speaker. On the off chance that you do not know how to do this, refer to the manual for your Bluetooth speaker.

- Open the Alexa application and go to Settings.

- Select the Echo Dot gadget and afterward go to Bluetooth > Pair a New Device. The Echo Dot will then enter pairing mode. Once Echo Dot finds your Bluetooth speaker, it shows up in the rundown of accessible gadgets in the application.

- Select your Bluetooth speaker. The Echo Dot registers with the speaker and then Alexa lets you know whether the connection is valid.

- In the Alexa application select Continue. The Bluetooth speaker is connected to the Echo Dot.

- On the off chance that you need to play sound from another gadget on your Bluetooth speaker, say "Disengage."

Tips:

In order to control sound volume when connected to a Bluetooth speaker, utilize the Echo Dot's volume control (found either by the light ring or by the voice) or the controls for volume on your outer speaker.

- To associate with a formerly combined gadget, say "Interface." Echo Dot interfaces with the last connected device.

- To deal with your Bluetooth gadgets, go to Settings > (Device Name) > Bluetooth.

How to Associate Echo Dot to External Speakers (Audio Out)

Utilize a sound link (excluded) to play sound from Echo Dot on different speakers.

Note: Playing music from another gadget on Echo Dot with a

sound link is not supported.

- Set your Echo Dot and outside speaker no less than three feet apart. If Echo Dot is too close to any outer speakers, it might be more troublesome for Alexa to hear the wake word or other commands.

- Turn your outer speaker on.

- Plug one end of a sound link into your speaker and the flip side into Echo Dot. All sound from Echo Dot will now stream through your outer speaker.

Note: If your speaker does not have a 3.5mm sound port, you can utilize a sound link connector.

Tip: To control the volume when associated with an outside speaker, utilize your Echo Dot volume controls (manual or voice) and the volume controls on your outer speaker.

Associate Echo to Wi-Fi

Your Echo requires a dynamic Wi-Fi association in order to follow your orders and stream media.

Before you start:

- Connect your Echo to an electrical plug.

- Open the Alexa application.

Echo associates with double band Wi-Fi (2.4 GHz or 5 GHz) networks that utilize the standard 802.11a/b/g/n. Echo does not associate with specially appointed (or shared) systems.

- In the Alexa application, open the left route board and then select Settings.

- Select your gadget and then select Update Wi-Fi. In case you are adding another gadget, select Set Up Another Gadget.

- On your Echo gadget, press and hold the Action button for five seconds. The light ring changes to orange and your cell phone connects with your Echo gadget. A rundown of accessible Wi-Fi systems will now show up in the application.

Note: The Alexa application may ask you to physically interface your gadget to your Echo through your Wi-Fi settings.

- Select your Wi-Fi provider and, if required, enter the system password. On the off chance that you do not see your Wi-Fi provider, look down and select Add a Network (for shrouded systems) or Rescan.

Macintosh system: If you have to add your Alexa gadget to your list of approved applications, look down on this screen until you see the MAC address.

Save your Wi-Fi password to Amazon

Any Wi-Fi passwords shared during set-up naturally show up when you interface another Alexa gadget to a similar Wi-Fi arrangement. The password is additionally required on the off chance that you switch between shared Wi-Fi systems. For more data, go to Saving Your Wi-Fi Passwords in Amazon FAQs.

Connect to an open system

If you are associating with an open system that requires a web program to sign in (such as a school or a hotel), enter any required data. The data could be a password, a room number, or button just an acknowledgement of the conditions for utilizing the system. This data can not be saved to Amazon.

Select Connect. After your gadget interfaces with your Wi-Fi network, a confirmation message shows up in the application. You are then prepared to utilize Alexa.

Reset Your Amazon Echo or Echo Dot (first Generation)

In the event that your Echo or Echo Dot (first Generation, with a volume ring) is inert, or you need to give your gadget to another person, you can reset the device. After you finish resetting your Echo Dot, you will need to re-register it again to an Amazon account and reset to default settings before using it again.

Tip: Having issues with your Echo? Restart your device first to check whether that resolves your issue. To restart your Echo gadget, unplug the power connector from the back of the gadget or from the divider outlet. At that point, re-connect the power connector.

To Reset Your Echo

Utilize a paper clip (or comparative device) to press and hold the Reset button (at the base of the device) for five seconds. After you press and hold the Reset button, the light ring around the top of your Echo Dot gadget will turn orange, followed by blue.

Sit tight for the light ring to go off and come on once more. The light ring will turn to an orange color again, and your Echo Dot enters into setup mode.

Open the Alexa application to network your gadget to a Wi-Fi system and register it to your Amazon account.

How to Reset Your Echo Dot (second Generation)

On the off chance that the Echo Dot (second Generation) is functioning poorly, or you need to give your gadget to another person, you have the option to reset the device. Once you have reset your Echo Dot, you will have to re-register it to your Amazon account and enter any device settings again before you can use it once more.

Note: Having issues with your Echo Dot? First, restart the device to check whether that settles your issue. To restart the Echo Dot, unplug the power connector from the back of the gadget or from the outlet in the wall. At that point, re-connect the power connector.

How to reset your Echo Dot (the device that has buttons for volume)

Press and hold the Volume down and Microphone off buttons simultaneously until the light ring becomes orange in color (around 20 seconds). After this, the light ring becomes blue.

Sit tight for the light ring to go off and come on once more. The light ring will appear orange once more, and then the set-up mode is activated on your Echo Dot.

Network your gadget to a Wi-Fi network by going into the Alexa application and registering the device to your Amazon account.

Chapter 5: How to Prepare Your Home for Amazon Echo Dot

Your Amazon Echo has quite recently arrived and you may want to just jump in and start using it. However, beyond the fundamental initial set-up, do you know what Alexa can do or what external programs you need to incorporate with your gadget?

Let's look at some of the top things that you need to do to begin.

Utilize the program interface

You will be instructed by the Amazon Echo Dot guide to download the Alexa application from Google Play or iTunes, yet a lot of individuals have had trouble while trying to use this method. We have been more successful utilizing the program interface.

In spite of the fact that the Alexa application is incredible for daily use, we suggest utilizing the browser for the underlying set-up, in case you have any issues with networking the Amazon

Echo Dot to a BT Home Hub. This also makes some other procedures, which we will look at in a minute, much easier.

How to Set Up a Default Music Service

Since a speaker is incorporated into the Echo, it is a decent Bluetooth speaker choice. The Amazon Echo is able to access your Amazon music library, which means any music you have obtained from Amazon, or on the off chance that you have a membership, Amazon Prime music.

In case you are not an Amazon Prime member, there is no compelling reason to stress. This is in light of the fact that Echo has TuneIn radio, a free addition, and also has the capacity to interface with Spotify.

Spotify also requires a membership. However, in the event that Spotifier is your standard music supplier then setting it up as an associate service is simple when you use the interface. Simply go to Settings > Music and Media and Snap to approve Spotify. This is when individuals have experienced issues, particularly on Android, so in case you are having issues, utilizing these steps may resolve matters.

Using basic commands, your Amazon Echo Dot is even capable

to read books to you. For instance, "Alexa, read me my book" initiates Audible or "Alexa, read my book on Kindle" will initiate Alexa perusing your Kindle library.

Changing Your Echo Streak Instructions

"Alexa, play my glimmer instructions" or "Alexa, what's new?" will begin your Amazon Echo newscast. As a matter of course, your Amazon Echo is set to utilize Sky News. But keep in mind that you may choose to get your news from an alternate source.

Under Settings > Streak preparation you can pick from a scope of various news administrations including BBC World Service, The Guardian, The Telegraph, and the sky is the limit from there.

Adjusting with your Google Calendar

Your Amazon Echo is entirely convenient at helping you to stay on top of your schedule. It's also ready to help you deal with your everyday exercises. By associating Echo with your Google Calendar, you can get Alexa to check what's going on today or at a particular hour on a day later on.

To do this just go to Settings > Logbook in the program interface. You will have to approve access to your date-book, but generally set-up is simple.

Acquiring More Echo Dot Abilities

Alexa is brilliant and there are a considerable number of things she can assist with. However, you can additionally enhance the convenience of your Amazon Echo by including significant aptitudes.

Amazon Echo Dot Skills

Aptitudes are basically modules that permit you to do an assortment of things, for example, requesting your last dinner from Just Eat or listening to UK radio. Simply enter the abilities area and scan for anything that fit your needs.

In our Amazon Echo Tips and Traps Article, you can discover a few different aptitudes that are fun to try out.

Associate Alexa to Smart Home Gadgets

Alexa is perfect with a scope of various Smart Home gadgets, giving you the ability to control everything from your heating to your lighting (assuming you have the required technological set-up in your home).

Initiate and associate the pertinent Smart Home abilities in the interface to begin.

Changing Your Echo Dot Wake Word

This one is not really fundamental, yet you may think that it's convenient, particularly on the off chance that you have somebody called "Alex" living in your home.

Under the gadget settings you can change the wake word to something else. However, at present your only choices are "Alexa", "Amazon" or "Echo". Oh, and no, there is no choice for "Hal."

Fun things to attempt

You have set up all the vital things by now. Why not experiment with something sillier and get Alexa to engage you? There is plenty to play with in this rundown.

- Alexa, what does WTF mean?

- Alexa, self-destruct

- Alexa, rap for me

- Alexa, Up, Down, Left Right, Left Right, B, A, Start

- Alexa, how much is that doggy in the window?

- Alexa, speak like Yoda

- Alexa, what is the Prime Directive?

- Alexa, I need reality

- Alexa, what does RTFM mean?

- Alexa, what would you like to be when you are all grown up?

- Alexa, do you know Hal?

- Alexa, when is the following full moon?

- Alexa, tell me something vague

- Alexa, what number of streets should a man stroll down?

- Alexa, what happens in the event that you cross the streams?

- Alexa, what does the fox say?

- Alexa, your mom was a hamster

- Alexa, what is war useful for?

- Alexa, do you have any new components?

- Alexa, Is Santa genuine?

- Alexa, INCONCEIVABLE!

- Alexa, you must be joking!

Meet the Alexa Family

Amazon Echo: The first $180 Alexa gadget has a vigorous speaker for playing music, and a great wide-exhibit amplifier that is continually tuning in. Say the wake word "Alexa", and even from an adjacent room the Echo will hear you and react.

Amazon Echo Dot: The $90 Dot is one of two new additions to the family. It's basically an Echo with the speaker slashed off, an additional sound yield port, and a Bluetooth radio. It's still continually tuning in, and it will even play music all alone. However, the sound quality is more similar to that of a cell phone. Because of the sound yield and Bluetooth, you can associate it to your own speakers or home entertainment framework.

Amazon Tap: The third individual from the Alexa family, the $130 Tap exchanges continual listening comfort for battery-fueled convenience. You need to hit a button to converse with the Tap, so you do not have to utilize the Alexa wake word. However, despite everything, you have access to all the same Alexa perks once you begin talking. The lightweight Tap can work like a regular, versatile Bluetooth speaker. Interface it to the web and it can stream music and let you utilize Alexa in a hurry.

Amazon has two other Alexa-related items:

Voice Remote for the Amazon Echo

A $30 Bluetooth remote that associates with your Echo or Echo Dot, permitting you to give orders beyond standard listening territory. You have to push the button on the remote to talk. It does not work with the Tap.

Fire TV Voice Remote: Included in the $50 Fire TV stick package or as a major aspect of the full-sized $85 Fire TV, it permits you to control your TV with Alexa with the push of a button, like the Tap or the Voice Remote. It only works with your TV, so I will not address this gadget while talking about how to best outfit your home with Alexa.

How Alexa Gadgets Cooperate

It's shockingly simple to add different Alexa gadgets to a solitary Amazon account. The Amazon Alexa application for iOS and Android walks you through the procedure. On the off chance that you synchronize control of any smart home gadgets to one

Echo unit, whatever other Echo items are attached to the same network will likewise remember them.

The Tap and the Dot make a convincing combo.

On account of a sound line-in on the Tap and the usefulness of Bluetooth, you can interface the Tap to the Dot either over a link or remotely. With the continually listening Alexa, this mix shows signs of improved Tap movability and better speakers. Purchasing both the Dot and Tap will cost $220, $40 more than a solitary Echo. However, it gives you a more adaptable set up with compactness when you require it—and you can depend on Alexa comfort when you do not.

Another approach to broadening your Alexa experience is a solitary Echo and a Voice Remote. That will cost $210 and it will give you the best speaker of the set, in addition to an Alexa contact point in one room. The Voice Remote uses Bluetooth and demonstrated broad range when we gave it a shot in the CNET Smart Home. I put an Echo on the top floor, went two stories down into the basement, and the Echo still reacted to my summons.

The Voice Remote works with either the Echo or the Dot.

The most moderate approach to amplifying Alexa's reach to various rooms is to purchase the Echo Dot and amplify it with the Remote. That setup will cost $120. This bodes well on the off chance that you already have a speaker framework for the Dot, or perhaps you couldn't care less about utilizing Alexa for music streaming.

Alexa Does not Sync

In spite of the fact that adding different gadgets to your Alexa application is consistent procedurally, you can not consolidate two Echo gadgets to make a stereo, or encompass sound set-up like you can with Sonos speakers. Indeed, putting two Alexa gadgets excessively near one another can cause problems.

On the off chance that you advise the Echo in your kitchen to play a tune, the Dot in your family room may likewise hear the command, and you will get both speakers playing a similar tune, without matching up. So, if you intend to use Alexa's continually listening mode, you need to select Alexa's location in your home carefully.

The rules below will help you determine how best to establish Alexa's range. However, in the event that you need to place continually listening Echoes close together, using the settings of the application you can program them with various wake words. "Alexa" is the most common. However, you can also utilize "Echo" or "Amazon."

This is what works with the Amazon Echo:

Suggested Set-up by House Estimate

In order to plan out a smart home set-up that works, the best thing you can do is to start with a couple of Alexa devices and figure out how your family and you will actually utilize them. Expand only once you are comfortable.

Smaller Homes, Lofts (one story; up to 1,000 square feet)

One continually listening Alexa gadget ought to cover it. Put either an Echo or an Echo Dot in a focal spot in the room you use most regularly, and you will rarely be out of voice range. Both the Echo and the Echo Dot are able to hear your voice through one closed door. Purchase a Voice Remote on the off chance that you have bunches of rooms, to minimize hollering.

Space constraints will make it hard for numerous individuals to issue commands to two Alexa gadgets without stepping on each other's toes. In the event that you require a speaker, utilize a Tap to minimize overlapping commands.

Medium Homes (one to two stories; 1,000 to 2,500 square feet)

More often than not, one Echo or Dot will be adequate if you put it in a focal room in a medium-sized home. If you want to give at least one of your children an Alexa gadget for playing music, utilize a Tap. You can also take a Tap to the parking lot or out to the yard, as you do chores. Utilize a Voice Remote for any fringe zones where you couldn't care less about music, yet still need to issue orders to Alexa.

In the event that you need broader general guidelines, plan to put two doors or one story between your Echoes or Dots to prevent overlapping orders. Those wide-exhibit receivers are extraordinary. However, they can also be baffling on the off chance that you have numerous relatives who need to play their own particular music simultaneously.

Vast Homes (two stories or more; more than 2,500 square feet)

As a rule, you will need to keep one Echo or Dot for each floor. In the event that you have a wide home, you ought to be fine setting up an Echo in your front room and a Dot in your family room as long as the length of a few hallways isolates the two spaces.

You may likewise need to strike a balance between the listening scope of the Echo or the Dot when it's peaceful and when you are playing music or you have the TV on. Alexa's outstanding listening range picks up on overlapping noise and the device can become confused. Similarly, the quality of Alexa's service deteriorates if the volume is too low.

The most ideal approach is with the Voice Remote. Put an Echo in the front room toward one side of your most-used floor. Place a Dot on the opposite side, then put a Voice Remote in any central rooms and you will generally have the capacity to give voice commands.

The Tap is a decent solution for any individual who needs an alternative music device in a room where this is already a continually listening gadget nearby. What's more, by and large, Taps are better for moving between garages, yards, and places where you just need to play music once in a while. However, it's not that much of a bother to unplug and replug the Echo or the Dot.

Chapter 6: How to Utilize Alexa Effectively

Amazon is continually adding new elements to its Echo gadgets.

With such a variety of voice commands that will work with Alexa, it's easy to miss a few of the less common ones. Since unpacking your Echo, you might've also missed some of the Echo's most underutilized features.

Here's a rundown of valuable tips that will help you get what you need from Alexa.

1. Make profiles for each relative

In the event that different individuals in your home have Amazon accounts, you can add them to your Alexa family unit profile. To check which profile is being utilized at present, just ask, "Alexa, which profile is this?" And to switch profiles, say, "Alexa, switch profiles."

While including extra individuals does not give you isolated shopping or schedules, it allows you to share content libraries. Be careful, however, as it also permits others to utilize your Prime account.

2. Change the wake word

Whether you have somebody in your home by the name of Alexa or you are worn out on Amazon's ads waking your speaker, changing the gadget's wake word is the best strategy.

Unfortunately, you are not yet able to set your own particular wake word. However, you can pick between the three options of Alexa, Amazon, or Echo. To choose one of the two exchange wake words, open the Alexa application or echo.amazon.com, open Settings, pick your Echo gadget, click Wake Word, pick one of the three choices from the dropdown menu and click Save.

3. 'Will you say that once more?'

If you ever miss one of Alexa's reactions or need her to rehash something, simply inquire. Say, "Alexa, would you be able to rehash that?" or, "Alexa, would you be able to state that once more?" She will repeat what she said as many times as you need her to.

You can also look into late reactions in the Home tab of the Alexa application, and listen to the recordings of the things you have said to Alexa.

4. Erase all your Amazon voice information

Everything you say to Alexa is recorded and transferred to Amazon's servers. If you do not like that many voice recordings of things you have said to your speaker are now gliding around in the Cloud, there are two approaches to get rid of those old recordings. Simply realize that even in the wake of erasing every one of the recordings, another one will be added each time you address Alexa.

5. Tackle fast transformations and math issues

Essentially, I utilize Alexa for controlling my Smart Home utilizing my voice. It's much easier than digging through my telephone to open several applications just to control the lights.

The second most helpful thing Alexa does for me is snappy calculations or conversions, particularly while cooking or making espresso. You can use her to switch currencies or to make estimations and even to do mathematic calculations. Simply say, "Alexa, 15 times 32" or "Alexa, 10 dollars to pounds."

6. Include Smart Home gadgets that aren't supported

In the event that you have a mix of Smart Home gadgets, odds are, there might be a couple that aren't formally supported by

Alexa. Be that as it may, you may not be stuck between a rock and a hard place.

In the first place, if there is not an Alexa skill for that smart gadget, check the online association destinations Yonomi and IFTTT to check whether your gadgets are supported. Provided that this is true, become more acquainted with Yonomi or IFTTT, as they can enormously grow the value of Alexa, permitting you to attach a few activities to a solitary voice command. It can also send your Alexa to-do list or grocery list to Apple Reminders or Todoist.

7. Stream any sound utilizing Bluetooth

As a matter of course, you can listen to Amazon Prime Music, Audible book recordings, iHeartRadio, Spotify, TuneIn Radio, and Pandora with Alexa speakers. You even have the choice of your Kindle books that you can have read to you by Alexa herself.

Be that as it may, in case you are hoping to play Apple Music, Google Play Music, your own music library or some other unsupported sound, you should first network the Alexa gadget to a PC or cell phone utilizing Bluetooth. At that point you can stream any sound you need through your Echo, Echo Dot or Tap, utilizing it as a Bluetooth speaker.

8. Set the default music administration to Spotify

As anyone might expect, Amazon gadgets consistently promote in-house platforms, for example, Kindle, Amazon Music, Audible etc. Be that as it may, with the Echo speakers, you can replace Amazon Music as the default music library with Spotify or replace the default station with Pandora or iHeartRadio. Simply open the Amazon Alexa application, go to Settings > Music and Media > Choose default music benefits and select your preferred applications.

What this means is that you need to be specific and need to indicate "on Spotify" each time you need to stream music. Rather, you should state "on Amazon Music" when selecting a song, to play it through Amazon's music platform.

9. Include Alexa Skills with your voice

Until recently, you could just add Skills to your Alexa gadgets by utilizing the Alexa application or echo.amazon.com to peruse or seek the Skills database and physically add them to your gadget. The procedure was inconvenient, fundamentally on the grounds that the Skills store was poorly networked.

Presently you can include Skills by voice. If you know the name of the Skill you need to include, simply say, "Alexa, empower LIFX" or "Alexa, empower Magic 8-Ball." After only a few moments, the ability will be empowered and accessible. Try it out with one of our top picks (duh) — the CNET News aptitude.

10. Turn on the wake-up sound

In case you are a long way from your Alexa speaker and can not see it, you might need to confirm that she heard the whole command. If you need to be sure that Alexa is tuning in, go to the Alexa application, open Settings > [Echo name] > Sounds and flip on the Wake up sound. For a confirmation that lets you know that Alexa heard your command, you can always empower the End Demand sound.

11. Include a Voice Remote

In the event that you have a tendency to sit too far from your Alexa speaker to dependably control it, for example, from the kitchen while the speaker lives in the family room, consider buying a Voice Remote for Amazon Echo. It's available for $30 or £20 and works for both the Amazon Echo and Echo Dot (not the Tap).

Synchronize the remote in the Settings menu in the Alexa application, and you can remotely converse with your Echo from over the house, in different rooms, or even while outside. It connects utilizing Bluetooth, so the range is restricted. In any case, it can spare you some disappointment in case you are not always close to your Echo.

12. Order anything from Amazon Prime

You can now order a great many items just by requesting them. Prior to this month, you could only reorder things you had obtained beforehand utilizing Prime. Presently you can order anything that is Prime-qualified and that is not attire, adornments, shoes, watches, Amazon Fresh, Amazon Prime Pantry, or Amazon Prime Now.

You can even request a Domino's pizza, or a ride from Uber or Lyft.

13. Track your Amazon purchases

In the event that you do happen to request something from Amazon with your Echo Dot, Echo, or Tap, you can track your purchases by asking, "Alexa, where is my order?" or saying,

"Alexa, track my purchase."

The data Alexa will give you about a package is not excessively detailed, yet it will let you know the day the packages are scheduled to arrive. What is needed now is an upgrade that will you know which service the package was mailed with (such as FedEx) so you can better calculate when it will arrive.

14. Get select Prime sales

Amazon likes to demonstrate its gratefulness to its most faithful clients, especially those who've spent real money on an Amazon gadget. Every now and then, Amazon will run exceptional sales that are held only for Echo, Echo Dot, and Tap proprietors. For example, the Echo Dot and Tap were at first accessible for purchase only to individuals who possessed an Amazon Echo.

The latest example was Prime Day, where various sales were for Amazon Alexa gadgets just - outsider Alexa gadgets and platforms like Echosim.io, Roger and Lexi were rejected from those arrangements.

15. Settle a conflict

On the off chance that you and a few companions are torn between where to go for supper or where you to go on vacation, you can place Alexa in control. Just command, "Alexa, flip a coin" to pick between two potential outcomes. You can likewise throw some dice or question a Magic 8-Ball (if you empower the aptitude).

The most effective method to create your own custom voice charges for Amazon Echo's Alexa

With a little assistance from IFTTT, you can show Alexa a wide range of new tricks. Here's how to get started.

From the beginning, we saw enormous potential for Amazon Echo to assume a focal part in people's homes. Soon Amazon opened up Alexa's product to outside designers and included support for savvy home rigging like Philips Hue's shading changing knobs and Belkin's WeMo line of brilliant switches and fittings. Extra outsider incorporations permit Alexa to arm your security framework as you are exiting the entryway, or even find your auto.

Amazon Echo has a channel on IFTTT, the free online platform that gives you a chance to make your own computerized "formulas" utilizing an "if this, then that" approach. That opens the way to utilize Echo with more smart devices and platforms.

And now, on account of a neat little overhaul to the IFTTT channel, you can utilize it to make your own Alexa voice commands for the Echo Dot to respond to, as well. Here are instructions on how to do that:

1. Interface Amazon Echo to your home's Wi-Fi arrange.

Obviously, you can not begin computerizing with Alexa unless your Amazon Echo is up and running on your home system.

Luckily, set-up is easy. Simply connect the Echo, network with its Wi-Fi platform on your Android or iOS gadget, and then open the Alexa application. From that point, select your Wi-Fi provider and enter its password to complete the process.

2. Activate the Alexa IFTTT channel.

To complete the steps to get your Amazon Echo Dot matched up with IFTTT, make a beeline for IFTTT's site and sign in (if you do not already have an account, create one. It's easy and free). Tap on "Channels" in the upper right corner, then look down and tap on the Amazon Alexa channel.

To activate the channel, you will basically need to tap on "Interface," then enter your Amazon log-in information to approve the combination. In the event that you'd rather do the

majority of this on your telephone or tablet, you can do so by downloading the free IFTTT application for Android and iOS gadgets.

3. Make your formula.

Now you can begin the process of creating new commands. Click "My Recipes" on the IFTTT landing page, then tap on the enormous "Make a Recipe" button.

IFTTT's formula creation instrument is entirely direct—you have to pick a trigger (the "if this" part) and an activity (the "then that" part). For our purposes, Amazon Echo will be the trigger. At whatever point we give Alexa a particular command, we need it to trigger a particular activity.

Tap the enormous blue "this" in the formula to begin. IFTTT will show a rundown of the greater part of its channels—tap on "Amazon Alexa" (you have officially activated it, so it ought to be ready). Once you have done so, you will see a rundown of triggers. You get a considerable measure of alternatives with Alexa, yet the one we need is "Say a Specific Phrase." Select it, and IFTTT will ask you what you need that expression to be.

A few quick things to note. You will have to state "Alexa" with a specific end goal in mind in order to wake your Echo up. Once it's listening for a command, you will have to state "trigger" keeping in mind that the end goal to instruct it to listen for a custom IFTTT express. Subsequently, every IFTTT command will begin with "Alexa trigger," and after will end with your custom expression. The "Alexa trigger" part is now accepted, so do exclude it with whatever expression you write in - on the off chance that you do, you will need to state it twice.

Your custom command should be composed solely in lowercase letters, and you ought to keep away from any kind of diacritical marks. For the time being, keep it basic and run with "my formula." Then wrap up by tapping on "Make Trigger." We're mostly done—our formula will fire each time we say "Alexa, trigger my formula." Now we simply need to make sense of the "then that" part.

4. Pick your activity.

Now, you will have to choose what it is that you really need Alexa to do. You will discover a lot of possibilities by perusing through IFTTT's developing rundown of channels, which incorporates online networking administrations, Web platforms, and plenty of interesting devices.

For this case, we will use the Nest Learning Thermostat. Amazon Echo is not a "Works with Nest" device, so it does not have a custom-manufactured indoor regulator mix of its own—and that is the thing that makes IFTTT so helpful. It can serve as an extension between brilliant Smart Home items that do not otherwise together.

The Nest channel gives me several choices: I can have Alexa set the warmth to a particular temperature, or I can modify the range between computerized warmth and ventilating. I can likewise trigger the HVAC fans to run for 15 minutes—that can prove useful, for instance, if I'm searing a steak in a cast iron skillet and know I may likely wind up filling my home with smoke during the cooking process.

When I say "Alexa, trigger my formula," I need the fans to kick on to get the smoke out of my kitchen. I should simply choose the "Turn on fans for 15 minutes" activity, then determine which Nest I need to control. (We have two of them set up in the CNET Smart Home, one for the upstairs and one for the ground floor).

Your alternatives will clearly differ based upon which activity channel you use, yet for every one of them, programming the desired outcome ought to be as simple as exploring a few drop-down menus and then clicking "Make Action." From there, IFTTT will offer a last rundown of your formula—if everything looks great, click "Make Recipe."

5. Continue testing.

As of now, you have got one custom command added to your repertoire. Why not attempt a couple of others? You have a lot of options, and you can even make an assortment of custom orders for a similar activity, keeping in mind that the end goal is to give your device use more flexibility. For example, saying, "Alexa, trigger the security framework on" and "Alexa, trigger the alert to outfitted" can both arm your IFTTT as a good security unit.

You can also always return to a current formula to change the verbiage. For example, if I decide that my "Alexa, trigger my formula" command was excessively bland, I can rapidly transform into something more particular and fitting, such as, "Alexa, trigger my fans since I simply set my kitchen ablaze."

The most effective method to set a repeating alarm with the Amazon Echo

The Amazon Echo Dot's alarm abilities have been missing since the device was included a year ago. That mistake has been rectified and you can now wake up to the sound of Alec Baldwin's voice each morning.

The Amazon Echo Dot makes quite an awesome alarm clock. Simply advise Alexa of what time you need to wake up.

Prior to the Echo Dot's latest significant redesign, you needed to let it know precisely when you needed to awaken each day. This wasn't particularly convenient.

Amazon has made life simpler of late by sneaking a repeating alarm into the Echo Dot's settings. Here's the way you can set an alarm that will repeat itself until you tell it to stop:

Setting the repeating alarm

To set any alarm with the Echo, you should utilize a voice command as the application and Web customer do not have a button to let you physically program in an alert.

When you are close to your Echo, say, "Alexa, set an alert." The Echo will ask when you need the alarm to go off. Answer so everyone can hear what the time will be. If you need the alert to repeat itself, determine how regularly this should happen. You can set the alert to go off daily, weekly, etc.

Alexa will confirm that you have set up the alarm and let you know when it's programmed to repeat.

Altering existing alarms

If you make a mistake while setting an alarm or have a current one you want to keep repeating, you can rapidly alter things.

To start with, open the Amazon Alexa application on your cell phone or go to echo.amazon.com from your program and explore to Alarms. Tap on the alert you have to alter. From that point, you can change the time and alert sound.

At one point, Alexa contained no options for selecting pick when an alarm repeated itself. However, now you can change that. Select one of the alternatives from the Repeats drop-down menu. When you have set up the alarms to your satisfaction, select Save Changes.

Dropping or erasing alerts

If you have to put an alert on hold or want to erase it through and through, you can do that from inside the application or Web customer.

Canceling out an alert essentially turns it off in the settings, yet does not erase it. To physically cancel an alarm open the Alexa application or echo.amazon.com, explore to Alarms and tap the button for the privilege to delay the alert.

You can likewise do this by saying, "Alexa, scratch off alert for [time and day]."

Erasing an alarm must be done from inside the application or Web customer. Explore to the Alarms menu, select the alarm you need to erase, and select Delete Alarm.

Alexa's Tricks and Tips

If you have one of these gadgets, you presumably know the nuts and bolts at this point. You can approach Alexa for the weather forecast for today, to search for daily newscasts, to program an oven timer, or to play a most loved playlist.

As the Echo group of items has expanded, so have the devices' abilities. Bodiless though Alexa might be, she can do significantly more than you may might suspect—everything from locating your lost phone to helping you get a decent night's rest. Here are nine approaches to maximizing your virtual, voice-fueled assistant:

1. Listen to podcasts

Alexa is most likely best known for its Smart Music application and can play tunes from an assortment of sources. However, it can also stream podcasts. That is because of the Echo's synchronicity with TuneIn, a platform that is home to live radio stations, as well as to a huge library of podcasts.

Hence, you can request that Alexa "play The Nerdist podcast" or "listen to The Truth on TuneIn". Despite the fact that you can interrupt and resume playback, you will lose your place if you request that Alexa do something else during that pause. If you need to listen to anything other than the track currently playing, you need to advise Alexa to "play the past scene"— and will have to do so again and again until you get to the track you need.

Obviously, that is not convenient. You can also seek TuneIn by means of the Alexa application, then pick the correct track you need to listen to. However, your sole voice-fueled alternative is to hear the most recent track.

2. Listen to the book you are perusing

If you invest a great deal of energy in your car or do a ton of walking, you do not need to worry about missing out on recent news or literature. Alexa can read the content of any book in your Kindle library. Indeed, it saves the spot where you stop each time, like an electronic bookmark, which is helpful at, say, bedtime or when you want to resume listening while throwing together supper in the kitchen. Simply say, "Alexa, read the Kindle book [Title of Book]."

Now for some bad news: Although Alexa is nice, it may not be the most enthusiastic of readings, nor can you, at this moment, substitute it for a well-rendered book on tape. Luckily, if you own a book recording variant from Audible, you can play that. Simply say: "Alexa, read the book recording [Title of Book]."

3. Change your default music service

When you request that Alexa play some music, it defaults to Amazon's own particular Prime Music library. That is fine if that is your usual go-to for tunes, yet if you prefer Spotify, iHeartRadio or Pandora, you may like to make that the default music app.

Luckily, it's anything but difficult to change the default setting:

- Open the Alexa application on your telephone or tablet.

- Tap the Menu button in the upper-left corner, then tap Settings.

- Look down to the Account segment and tap Music and Media.

- Tap Choose default music administrations. Under Default music library, tap the drop-down music menu and pick Amazon Music or Spotify (you can also go under the Default station where you can select Pandora, Amazon Music, or iHeartRadio). At that point, tap Done twice.

You can also make Spotify, Amazon Music, Pandora or iHeartRadio your default music app.

Obviously, whether or not any of these apps are your defaults, you can play music from any of these platforms simply by issuing a specific verbal command. You should simply say, "Alexa, play the Holiday Classics playlist on Amazon Music."

Take note that Alexa does not, as of now, work with Google Play Music or iTunes/Apple Music. To play music from both of those sources, you can do one of two things: Use your PC to transfer your list to Amazon Music or else Bluetooth connect your phone to your Echo Dot, Echo or Tap and stream the music by means of Bluetooth.

4. Get talented at including skills

Given everything Alexa can do, it's hard to grasp all of its vast capabilities – known in Alexa jargon as "aptitudes." (Think apps, specific to Alexa.) There are a large number of them, and you can include any one just by saying, "Alexa, empower the X application."

To illuminate for X, make a beeline for the Alexa application on your cell phone, tap the Menu button, and then tap Skills. You can peruse the different categories or look for something else you might need. For instance, in case you are hoping to enhance your vocabulary, empower the Daily Buzzword ability. Consequently, when you say, "Alexa, open my Daily Buzzword," and she will give you another word and it's meaning from the dictionary, finishing with a proprietary test.

Alexa has abilities for pretty much all that you can envision, from home robotization to care reflection to significantly essential feline truths application. Given that the commands need to be word-specific, keeping a paper list close to your Echo can be helpful.

5. Match it with IFTTT

On the off chance that you need to influence outsider instruments and administrations that do not yet have going with abilities, look to IFTTT. For those new to it, IFTTT (another way to say "If This Then That") is a free administration that utilizes "applets" (previously formulas) to interface with different administrations. You can make these applets yourself or utilize existing ones.

Looking at the site for "Alexa" or "Echo," you will discover many applets you can begin utilizing instantly. For instance, do you like utilizing Alexa to add things to your schedule? In case you are an iPhone owner, download the iOS Reminders application. Do you regularly utilize your Echo to set clocks? There is an applet that can send a warning to your telephone when a clock goes off.

These are just the tip of the iceberg of Alexa's abilities. If you have ever wished that Alexa could sync up with application specific app or platform, IFTTT can probably make that wish come true.

6. Get some rest

Alexa gadgets make awesome end table sidekicks, beginning with the alert component: "Alexa, set an alert for 6:15 a.m." Here's something significantly cooler: If you wander into the Alexa application's settings, you can alter your alerts so that celebrities offer your wake-up calls (Alec Baldwin, anybody?).

Simply do the following:

- Open the Alexa application on your telephone or tablet.

- Tap the Menu button, then Timers and Alarms > Alarms > Manage alert volume and default sound.

Tap Alarm > Celebrity and pick the voice you need.

Even better, Alexa can play a wide range of sounds to help you nod off: book recordings (anything from your current Audible library), podcasts (by means of Tuneln), and even background noise type of "resting sounds" from Amazon Prime Music. Once you have started your music playing, you can then train Alexa to "set a rest clock for X minutes," so you do not need to wake up to instruct it to turn the music of.

7. Discover your telephone

You are five minutes late getting out the door; you unquestionably do not have sufficient energy to chase down your telephone. Did you leave in the room? Restroom? Did it slip underneath the couch?

Alexa does not know the reply, however, it can help you chase that missing item down, so long as your device is properly equipped. Begin by downloading the TrackR application for iOS or Android onto your telephone; it's a free locator application that has a buddy Alexa application. Once you have downloaded the application and included the ability, you can state, "Alexa, where's my telephone?" and Trackr will make the telephone ring boisterously, even if it's in silent mode.

You can likewise set up an IFTTT "formula" appropriately titled, "Advise Alexa to discover your telephone." For this situation, Alexa simply dials your telephone—not the best choice if it's normally set to silent.

8. Control your home

Alexa can "talk" to a developing number of Smart Home gadgets, everything from lights to electrical outlets to thermostats. You can now control all these gadgets using apps on your telephone or tablet. However, it's much faster to issue a verbal command than it is to scroll through a list of phone apps.

Maybe you left the remote control on the first floor. Just say, "Alexa, kill the TV." Or you are walking in the door at dusk. Say, "Alexa, turn on every one of the lights." Can not remember if you locked your doors? The command is obvious: "Alexa, look my doors." Even cooler, you can make bunches (inside the Alexa application) to control multiple devices simultaneously.

If you are like me and you do not have an incredible memory, Alexa can help there too, by allowing you to give your devices multiple names. For instance, if you forget whether you called a light the "lair light" or "light in the sanctum," simply utilize "cave light" as the essential name of the gadget and afterward make a gathering called "light in the nook" with your cave light as a part. Either command will allow you to control the light. (Thanks to Computerworld's Sharon Machlis for this tip.)

Obviously, this is all dependent upon the devices you purchase the switches and apparatuses you need to control. Before you buy anything smart, check to ensure its Alexa-compatible.

9. Simply have a great time

Alexa has a sense of humor, as confirmed by its reactions to inquiries and expressions like these (all introduced by "Alexa," obviously):

"I wasn't expecting the Spanish Inquisition."

"Will reindeer fly?"

"Why so genuine?"

"Let me know a tongue twister."

"Where would I be able to conceal a body?"

"What is the significance of life?"

"Up, Down, Left Right, Left Right, B, A, Start."

Alexa is truly proficient at popular culture references as well. Just for fun, check out how it reacts to expressions like, "Winter is coming, "or "These aren't the Droids you are searching for."

Top Alexa Skills for the Amazon Echo Dot

According to Amazon, a great many Alexa-empowered gadgets were sold this Christmas season. More than 5 million Alexa-empowered gadgets were actually sold *before* the start of the Christmas season.

Which aptitudes have risen as the year's top picks? The research directs to Amazon's 2016 Alexa Skills Customer Picks and at Alexa Skills Marketplace for the most-empowered abilities.

Thunderstorm Sounds

Alexa has such a variety of capacities that it's difficult to remember sometimes that its simplest functions may be its most popular (and useful).

Music is the No. 1 reason individuals utilize Alexa, and soothing sounds like electrical storms, rain, and sea waves—are ranked at the top of popularity charts.

Jeopardy! J6

This is a day-by-day gathering of six trivia questions asked in a Jeopardy! style. Like the show, your response must be stated as a question, and you have just a few moments to react. Every question was created by Jeopardy! show writers.

Luckily, J6 trivia is simpler than trivia from the show.

The Magic Door

This is a pick your-own-experience application that gives you a chance to do things like investigate the ocean, slopes, or forest.

There are resting mythical serpents, sanctuaries, mood music, and different baffling sights and sounds to lure you in. This can feel like a superfluous device function, but try this with other people and it's immediately more fun. Kids truly like it.

I wish this ability gave you the choice to begin and stop. In the event that you stop whenever, the experience ends.

Stories aren't intended to be told in Alexa's monotone voice, yet this aptitude has proved highly popular application. It gets four out of five stars and a major shout-out from the Alexa client group.

Short Bedtime Story

Each of these stories is around 30 seconds to one minute long and is personalized to your child's name.

There is a Minions-related story and one that happens "in a world far, far away." There is one in which a child president gets chosen for a one-minute term for giving everyone dessert. You get the idea.

One story is about a zombie assault, which may not be precisely bedtime material for most children who are still young enough to need that kind of thing to lull them to sleep.

In one story, my girl was named a Nobel Prize champ. I like that. She likewise got a $100 Amazon Prime blessing declaration. I'm not so fond of that. That is the main element of commercialization I found in the stories.

When I played this skill for my child, she requested increasingly difficult stories, so that was promising.

This aptitude ought to permit more customization — so you can choose the age of the tyke, not only their name — on the grounds that the sort of story a five-or seven-year-old needs to hear is not necessarily appropriate for, say, a three-year-old.

On account of Alexa's monotone voice, I do not think I'd ever make her my go-to bedtime stories, but the function is fun to play around with.

Ask My Buddy

Ask My Buddy is a personal readiness platform. It will not call 911, yet in the event that you require help, you can advise Ask My Buddy to alert one or more of your friends, relatives, or bosses via text, email, or even a telephone call.

This aptitude takes advantage of how convenient hands-free devices are. It might explain why Echo or FireTV can be found in the homes of senior subjects who might not generally be inclined to seek out an AI assistant.

The Alexa programming advancement pack has as of now been utilized to place Alexa into a wheelchair. Years from now we may all think back and ask why we ever constructed a stick without Alexa worked in.

This application originates from the 2016 Alexa Customers Picks list.

It is perfect for hands-free gadgets, but perhaps less for owners of the Tap Echo.

20 Questions

This resembles trivia turned on its head—you have to find a

question instead of an answer—, and it's entirely brilliant.

One reviews: The questions can come too quick to permit you to think. Slowing things down would make the game more fun.

Inspire Me

This application gives you a sound cue — finishing with moving music — like words from Steve Job's popular 2006 commencement speech at Stanford University.

I could see this being fused into an IFTTT applet to begin your day.

Applets are what IFTTT used to call "formulas," and you make them to consolidate various commands into a solitary voice-empowered order for devices such as Alexa and Google Assistant.

So, you can create an applet that is activated when you say "Alexa, it's a great opportunity to begin the day" that will turn on your lights, begin making your espresso, give you a newscast, and then start the Inspire Me aptitude.

I can understand why this ability is popular, but it is missing many highly popular "inspirational" figures. Instead you get competitors, big names, and "thought pioneers."

I requested Gandhi and got Tony Robbins. I requested Tupac Shakur and Alexa came back to me with Oprah Winfrey. Confucius returned with Peyton Manning. Where's Kennedy? Churchill? Frida Kahlo?

This is a cool idea, yet the ability needs more quotes from genuinely rousing individuals to be viewed as complete.

Maybe the best part of this aptitude: Each inspirational quote delivers information about the speaker to your Alexa application, so you have the opportunity to learn more about them, if you are interested.

My Pregnancy from BabyCenter

This application can give you week-by-week updates about what's going on with your pregnancy. A large number of parents depend on BabyCenter, so the application and Alexa's synchronicity with the app works well.

The Bartender

This application teaches you to prepare more than 12,000 distinctive mixed drinks. It's not exclusive to Alexa. It's one of the main apps available from Google Home.

Fitbit

Together with Alexa, Fitbit was a standout amongst the most-downloaded free applications this Christmas. The application can let you know more than twelve things about your wellbeing and movement, similar to how much rest you got the previous evening or your resting heart rate.

As the Fitbit site clarifies, it can not help you log hydration or caloric intake. I've never understood why such a large number of wearable gadgets give information without explaining what the data means. All right, you made 10,000 strides today and burned a particular number of calories, but so what? How does that compare with seven days back? In particular, have I achieved my fitness goals? Should a man my age, height and weight accomplish more?

Individuals require a mentor, not a spreadsheet.

Alexa's Easter Eggs

Alexa, Amazon's virtual assistant, has the capabilities to stream music, read to you the day's features, or help manage the devices in your Smart Home. However, 50% of the fun is making inquiries out of left field to check whether she can keep up.

Surprisingly Alexa adores both a decent inside joke and popular culture. The ability to riff off your Wars reference or to avoid inquiries concerning where babies originate from is all part of Alexa's appeal. Without it, the Amazon Echo would be a less pleasant AI. Nobody likes conversing with a stuffy, withdrawn, know-it-all.

Alexa's Cloud-controlled smarts are stuffed with Easter eggs, and Amazon updates them consistently. Here are some of our current favorites:

Alexa's nerd cred

Think about your most loved nerdy show or film, and there is a decent shot that Alexa's a fan. Trekkers will welcome the following Alexa commands:

"Alexa, beam me up."

"Alexa, set phasers to execute."

"Alexa. Tea. Earl Gray. Hot."

The Star Trek references are the tip of the iceberg. Alexa will react to a few great Star Wars quotes, She will clarify the hypothetical outcomes of the intersection of the streams. Fans of the hit show "Game of Thrones" can say to Alexa "Winter is coming" and they will hear one of a few quotes from the show. In case you are a devotee of "The Big Bang Theory," she will even take you on in a round of rock, paper, scissors, reptile, Spock.

"Alexa, start self-destruct grouping."

Alexa is strangely enamored with any sci-fi that involves AI brainpower gone awry. That includes movies classics like "2001: A Space Odyssey" ("Alexa, open the unit narrows entryways") and "Eliminator" ("Alexa, are you Skynet?"), alongside darker alternatives, similar to 1983's "War Games" ("Alexa, I need to play worldwide atomic war.")

In the event that you ever start to worry that Alexa is getting too smart for her own good, you can simply request that she start her self-destruct succession.

Alexa's knowledge is not restricted to science fiction, however. Experiment with any of the accompanying film references, and Alexa will know precisely what you are discussing:

"Alexa, my name is Inigo Montoya."

"Alexa, I need a reality check."

"Alexa, party on, Wayne."

"Alexa, show me the money."

"Alexa, what's the first (or second) administer of Fight Club?"

"Alexa, without a doubt you must be joking."

Some of my top picks are Alexa's Monty Python references. You can, for example, approach her for the air speed velocity of an unladen swallow, and after that follow up by asking how she knows such a great deal about swallows. Oh and, you could get to the point and advise her that her mom was a hamster.

Games

Alexa is gamer-accommodating, as well. Besides Alexa's abilities to recall Minecraft formulas for to let you know what covering is available to be purchased for the current week in Destiny, Alexa is capable of a few exemplary gaming references.

For example, in case you are a devotee of the Portal arrangement, you could take a stab at inquiring as to whether the cake is a lie, or ask whether she knows GladOS (yet another insidious AI with whom Alexa has an unsettling kinship).

If you have played through the Mass Effect arrangement, you can take a stab at asking Alexa, "Does this unit have a spirit?"

For something somewhat more high level, try advising Alexa to do a barrel roll. Or, if you have never deceived your way through Contra, utilize the Konami code on her (that is Up, Up, Down, Down, Left, Right, Left, Right, B, A, Start).

After featuring in a Super Bowl commercial for Amazon Echo, Dan Marino, Hall of Fame NFL quarterback, and Alec Baldwin have both shown up in cameo appearances with Alexa.

Superstar cameos

Superstars are beginning to get in on the fun, as well. Taking after their appearance in Amazon's first-since forever Super Bowl advertisement, you will hear visitor cameos from Alec Baldwin and Dan Marino, Hall of Fame NFL quarterback. Try asking Alexa the amount of Oscars that Alec Baldwin has been awarded or how many rings from the Super Bowl Dan Marino has. (The reply in both cases is "zero.")

Baldwin and Marino are the main voices, other than Alexa's, that you can hear by asking Echo a question. However, as the device becomes more and more popular, expect to see different stars take action accordingly with Easter eggs of their own.

All the correct answers

Not each Easter egg is a conspicuous popular culture reference or a play on a film star's sense of self. In some cases, Alexa will offer shockingly spot-on answers to intense inquiries, or nuanced avoidances to inquiries you'd prefer she ignore.

This may prove to be useful if you have youngsters. On the off chance that they get some information about Santa Claus, Alexa will play along and enlighten them with incredible things concerning the jolly old man. Ask her how infants are made, and she will stammer out a cumbersome answer suggesting you go ask Mommy or Daddy instead of her.

Alexa is well-versed in the sorts of logical inquiries that you would except any AI to be acquainted with. Give her some questions about the laws of mechanical technology, and she will give you a pitch idealize rundown. Get some information about the Turing test (and regardless of whether she passes it), she will clarify that she does not have to pass it since she is not attempting to become human. No word yet on whether she could beat you in a session of Go, however.

Like Alexa's different Skills, the array of Easter eggs will probably continue developing.

Chapter 7: Troubleshooting the Amazon Echo Dot

The Amazon Echo Dot speaker is probably going to end up a standout among Christmas presents for techies who want to streamline their home existence. Though our experiences with Echo have been enjoyable thus far, here are some issue you might encounter, along with some potential solutions:

1) THE WI-FI CONNECTION IS INCONSISTENT OR NON-EXISTENT

Alexa's availability status is shown by the power LED on the base of the gadget, with white meaning great and orange meaning no Wi-Fi network.

In case you are encountering constant network interruptions while using your Wi-Fi with Amazon Echo, this is what to attempt.

To begin with, try the standard method: reboot your Echo, then reboot it once more.

If that does not work, you might need to consider repositioning your Echo further away from other devices that could be interfering electronically.

You can also reduce service interruptions by deleting unused apps from the system.

If you have a double band modem, you should be able to adequately have two systems set up. Try changing the device from 2.4GHz recurrence to 5GHz recurrence or the other way around. 5GHz guarantees less interference, better speeds, and a more reliable network.

On the other hand, 2.4GHz can be better for devices that are far from the Echo, particularly if the signal needs to be transmitted through closed doors.

Move your Echo to higher ground, like a bookshelf, to minimize outside interference.

2) ALEXA WILL NOT CONNECT TO OTHER DEVICES

One of Alexa's gifts is her capacity to control Smart Home devices such as Philips, SmartThings, Honeywell, Wink and Insteon.

First, ensure your device is properly synchronized with the echo. It might require a scaffold like the Wink center.

The following step, of course, should only be taken after the set-up guidelines for the Smart Home appliance have been followed. Otherwise, Alexa will be unable to find the device. You may need to download a partner application effort to assist you.

This may go without saying, but ensure the Smart Home devices are networked with the same Wi-Fi network that the Echo is on. If you notice that your devices have a tendency to default to the BT Open zone hotspot, this could be problematic.

Make sure you have downloaded the latest firmware and updates for your device, and remember that many Smart Home gadgets require you to empower them as an "ability" in the Alexa application before they are function.

Open the application, hit the menu and select "Abilities" then look or scan for the manufacturer. Different gadgets, similar to Philips Hue, do not require an "Aptitude" and can be connected quite easily by requesting that Alexa "Find gadgets" while squeezing the button on the Philips Hue connect.

The Alexa application empowers you to connect Smart Home gadgets in a Group. Along these lines, you can utilize voice controls to control various gadgets with a solitary command. For example, "kill room lights."

If Alexa is not picking up your command, it might be because she does not understand the wording. Speak simply and clearly.

Keep in mind, if Alexa does not support your preferred gadget, you can simply make an IFTTT formula to get around it.

3) ALEXA DOES NOT UNDERSTAND ME

One of Alexa's most used phrases is that she is sorry, she does not understand the question that you have asked. It can be truly disappointing. The more you use her, the more Alexa will train herself to learn your vocal nuances. However, while that process is ongoing, there are ways to avoid having to repeat your commands over and over.

Begin by utilizing the voice preparation device. In the Alexa application, go to Settings > Voice training. You will be asked to speak 25 pre-selected expressions to assist Alexa with learning your voice.

Next, determine what Alexa really heard. The Alexa application makes note of all your commands, so you can check and find out precisely what she heard. You need to go to the application's Settings and look at the History. In History, you can determine which words are normally misheard and therefore express them more clearly next time.

Take note of Alexa's position. Have you placed her near loud appliances such as the TV, cooling vents, dishwasher or stereo? Amazon says that baby monitors and microwaves can cause interference. They advise placing the Echo no less than eight inches from any wall.

4) MY BLUETOOTH DEVICES AREN'T CONNECTING TO ALEXA

The Echo Dot underpins the Audio/Video Remote Control Profile (AVRCP) and Advanced Audio Distribution Profile (A2DP SNK). Therefore, make sure the gadget you are endeavoring to network matches up. After this, the next step, obviously, is to make sure that your Bluetooth gadget is charged.

In case you are still encountering issues, you can try to un-pair and re-match your Bluetooth gadgets. You need to go into the Alexa application and go to Settings. Then, tap on the Echo gadget, highlight Bluetooth and Clear all combined gadgets.

To re-match the gadgets, say "Pair" to Alexa to put the Echo Dot in discovery mode. After this, go to the Bluetooth settings on the gadget or application. Alexa will confirm the association.

5) STREAMING SERVICES AREN'T ... STREAMING

One of the best skills that the Echo has is its capacity to stream media from various sources, including Spotify, TuneIn, Pandora and iHeartRadio. In case you are encountering interference, it's conceivably down to Wi-Fi obstruction. If that is the issue, follow the steps mentioned earlier.

Other potential interferences with Alexa's signal are your web speed and any firewalls you have on your system.

If your network is under 0.5Mbps, you are likely stuck between a rock and a hard place with regards to compelling spilling, while Amazon encourages you to guarantee the accompanying ports are open in case you are running extra safety efforts: 123, 443, 4070, 5353, 40317, 49317, 33434.

6) CHANGING THE WAKE WORD

On WWE Smack down, there is another character called Alexa Bliss, who continues waking our Echo. You can change the wake word in the Alexa application by picking Settings > Your Echo > Change wake word. Lamentably, the only choices are the fairly un-fun "Amazon" and "Echo".

7) IF ALL ELSE FAILS...

This ought to be a final resort. If all else fails, you can do a factory reset for your Amazon Echo and start all over again with its programming. There is a reset button beside the power connector. You will have to utilize a stick (or something similarly little and pointy). Hold the button in until the light ring turns orange and then you are ready. You will need to set up the Echo once more after this.

Step by step instructions to prevent Alexa from requesting stuff without your consent

Recently, news got around about a child making a substantial purchase on Amazon on account of some assistance from Alexa. The story was reported by a San Diego-based morning news program, which described the young lady and her mom saying, "Alexa, get me a dollhouse," in addition to other things. This incited quite a few instances of coincidental Alexa triggers over the San Diego range.

I viewed the news clip myself, and beyond any doubt, Alexa on my Amazon Dot thought I was attempting to place an order for

Amazon Echo Dot

a dollhouse ... twice. Try not to stress, however. Alexa is set up to request that you confirm any purchase before proceeding. Therefore, when Alexa heard the news report and thought I needed to purchase a dollhouse, it let me know what the most well-known one on Amazon was and inquired as to whether I needed to get it. It did not just automatically buy the item.

That does not mean that Alexa might not one day accidentally overhear your TV, think it's making a purchase, and then hear an additional 'yes,' from the television, causing it confirm the sale.

There are two or three ways you can prevent Alexa from requesting stuff without your consent, which is particularly helpful if you have someone in your home (like a creative kid) who likes to buy things without asking first.

Step by step instructions to disable shopping through Alexa

If you do not want to use Alexa for shopping, you can disable the function.

- Open the Alexa application.

- Tap the Menu tab in the upper left corner.

- Tap Settings.

- Look down and tap Voice Purchasing.

- Flip the switch for voice purchase off.

- Tap Voice Purchasing, then kill Voice Purchasing

Whenever this option is disabled, nobody will be able to buy things from Amazon utilizing Alexa on Echo or Dot.

Step by step instructions to add a password for purchasing confirmation

If you want to use Alexa to shop online but also want a safety net to ensure unwanted purchases do not occur, you can set up a four-digit password.

- Open the Alexa application.

- Tap the Menu tab in the upper left corner.

- Tap Settings.

- Look down and tap Voice Purchasing.

- Enter a four-digit password into the confirmation code box.

- Tap Save Changes.

After you have set a password up, every time you decide to

purchase an item, Alexa will state, "Let me know your voice code." Without that code, you can not finish placing the order.

Chapter 8: Recommendations

The Amazon Echo Dot is clearly a very useful little device as evidenced by the fact that after the Thanksgiving sales, not a single speaker was available online. Therefore, it is only natural that you want to make the most of your Echo Dot device in terms of enhancing its performance and ensuring that you get the most out of it.

In order to do so, you may have to purchase some additional products to improve your Amazon Echo Dot's performance. Below are some products that are a good idea to purchase along with your Echo Dot.

Amazon Echo Dot Case (fits Echo Dot 2nd Generation only)

You no longer have to limit the color of your Amazon Echo Dot to boring black or withering white. Match your device to the furnishings in your room by using the Amazon Echo Dot Case. Please keep in mind that this is meant for the 2nd Generation devices only. You have a range of six colors to choose from— charcoal fabric, merlot leather, indigo fabric, saddle tan leather, Midnight leather and sandstone fabric. The fabric options are made of nylon woven fabric while the leather options utilize full grain leather. These cases are made to fit precisely around your device but do not hinder access to any of the ports or buttons. They also allow you to see the light ring from any angle.

Alexa Echo Dot Voice Remote (Works with Amazon Echo and Amazon Dot)

The voice remote allows you to communicate with Alexa when you are too far of a distance from the speaker itself. There are buttons on the remote that are used for volume control, next, previous, and a play/pause function. The dimensions for this device are 1 inch in width and about 4.5 inches long. It is about twice the size of the Fire Stick. While this is a handy and convenient tool, keep in mind that if you have more than one Echo or Dot, you will need to purchase additional remotes because once paired to one device, it cannot be used on the others. It sells on Amazon for $29.99 and can be found as a Prime item, giving you free 2-Day shipping.

Portable Battery Base For Echo and Echo Dot

This portable battery base will allow you to take Alexa anywhere. It allows for up to 6 hours of continuous listening and comes with a soft silicone pad for added protection. It fits securely with the base of the Echo and easily attaches. With this device, you are able to leave the power adapter plugged into the battery base and due to the Intelligent Battery Management System it will not damage the battery.

The battery contains three light indicators for On/Off status, Low Battery notification, and Charger Status. You can order these on Amazon for $49.99 as a Prime item with free 2-Day shipping.

Leather Cover Sleeve Skins for the Echo and the Echo Dot

Leather cover sleeve skins are a great accessory to protect your device for surface scratches, while giving your Echo a more personalized feel. They come in a variety of colors and designs and come with an attachable strap for convenient portability. The sleeve contains vent openings that will not hinder the sound quality coming from the speakers. You can find a variety of these on Amazon for as low as $12.99 and are also a Prime item with free 2-Day shipping.

Acrylic Speaker Stand for Echo and Echo Dot

If you live in a home with pets or small children, this accessory is a must have. It is constructed to allow optimum sound quality while providing stability against tips or falls. It has a horseshoe design to allow for easy placement and blends in well with the sleek black design of Alexa. You can find this as a Prime item on Amazon with free 2-Day shipping for $24.99.

Wall Mount Holder for Echo Dot (2nd Generation)

This wall mount secures tightly to a wall or flat surface to securely hold your Dot in place. There are hook clips to ensure the security of your Dot when hanging on a wall and is constructed of clear plastic with venting to allow optimum usage of voice controls. These handy accessories can be found for $12.99 as a Prime item on Amazon with free 2-Day shipping. You can also find some that have hooks so that you can move them room to room wherever you go and they can easily hang on a preinstalled hook.

2-Year Protection Plan plus Accident Protection for Echo Dot

The Amazon Echo Dot is an electronic device and like many other such devices, it is susceptible to accidents and damage from such accidents. You need to ensure that in case of such accidents, you do not end up paying through your nose while trying to get your device fixed. This is where the 2 Year Protection Plan plus Accident Plan for Echo Dot comes in. The plan is available for 2nd Generation Echo Dot purchased within the last thirty days and is delivered via email, which means that you do not have to go through the extended hassle of paperwork.

Your card is charged as soon as you make the purchase and the protection begins as soon as the device is delivered to you. You can make up to three claims during the warranty period and you will get a replacement device within two to three days after you have made a claim that is determined to be valid. You can cancel it any time and get the full refund if you do so within the first 90 days of the service. There are no deductibles or shipping fees (it's just an email, after all) and if you are buying the device as a gift for someone, you can transfer this protection fully to that device too.

799 Funniest Things to Ask Alexa (Amazon Alexa, Echo Dot, Amazon Echo, Alexa): Kindle Edition

Yes, we have already mentioned the fact that Alexa can have a sense of humor and is full of Easter eggs to delight the heart of everyone who has ever given her a frivolous command. However, are you sure you have exhausted all the inane, geeky or just plain silly commands that you can give her? Our recommendation is this particular eBook that gives you an exhaustive list of almost eight hundred commands—some of which will be familiar from what you have read earlier in this book.

There are hundreds of other commands just waiting for you to issue them. Use these commands as a party trick or amuse your friends with them or just issue them to get some laughs at the end of a stressful day.

Chill out with *799 Funniest Things to Ask Alexa.*

Conclusion

And here we are. As all things must, this book too must end now. I hope that it gave you effective insights into all things Amazon Echo Dot and Alexa related.

We took a look at what voice assistants are and then examined the Amazon Echo and Echo Dot. We discussed Alexa in great detail and discussed how to get the most out of her including how to increase her skills and even Easter Eggs. We examined common problems faced by Amazon Echo and Echo Dot users and the solutions or workarounds for these problems. We also took a look at some of the products you can purchase along with your Echo Dot to ensure it stays safe and protected and looks good while doing so.

Do leave a review to let me know what you thought of this book and what, if anything, you would like me to add to it.

Works Cited

Broida, R. (2017, January 20). *9 Alexa Tips and Tricks: Amazon's Echo, Echo Dot and Tap devices can do more than just check the weather and sling playlists.* Retrieved from Computer World: http://www.computerworld.com/article/3153032/home-tech/9-alexa-tips-and-tricks.html

Crist, R. (2017, January 16). *How to craft your own custom voice commands for Amazon Echo's Alexa.* Retrieved from www.cnet.com: https://www.cnet.com/au/how-to/how-to-craft-your-own-custom-voice-commands-for-amazon-echos-alexa/

Crist, R. (2017, January 16). *The weird, wide world of Amazon's Alexa Easter eggs.* Retrieved from Cnet: https://www.cnet.com/news/amazon-echo-alexa-easter-eggs/

Johnson, K. (2017, January 9). *10 Alexa skills for your new Echo, ranked by popularity.* Retrieved from Venture Beat: http://venturebeat.com/2016/12/27/top-10-amazon-echo-alexa-skills-2016/

Tew, S. (2017, January 6). *15 Amazon Echo Tips and Tricks.* Retrieved from www.cnet.com: https://www.cnet.com/how-to/amazon-echo-alexa-tips/

www.ingramcontent.com/pod-product-compliance
Lightning Source LLC
Chambersburg PA
CBHW051810170526
45167CB00005B/1956